Protein NMR Spectroscopy:
Practical Techniques and Applications

Protein NMR Spectroscopy: Practical Techniques and Applications

Edited by

Lu-Yun Lian

NMR Centre for Structural Biology, Institute of Integrative Biology, The University of Liverpool, Liverpool, UK

Gordon Roberts

Henry Wellcome Laboratories of Structural Biology Department of Biochemistry, University of Leicester, Leicester, UK

WILEY

A John Wiley & Sons, Ltd., Publication

This edition first published 2011

Registered office
John Wiley & Sons Ltd, The Atrium, Southern Gate, Chichester, West Sussex, PO19 8SQ, United Kingdom

For details of our global editorial offices, for customer services and for information about how to apply for permission to reuse the copyright material in this book please see our website at www.wiley.com.

Library of Congress Cataloging-in-Publication Data

Protein NMR spectroscopy : practical techniques and applications / edited by
Lu-Yun Lian, Gordon Roberts.
p. cm.
Includes bibliographical references and index.
ISBN 978-0-470-72193-3
1. Proteins–Analysis. 2. Nuclear magnetic resonance spectroscopy. I.
Lian, Lu-Yun. II. Roberts, G. C. K. (Gordon Carl Kenmure)
QP551.P69725 2011
547′.7–dc22

2011010948

A catalogue record for this book is available from the British Library.

Print ISBN: 9780470721933
ePDF ISBN: 9781119972013
oBook ISBN: 9781119972006
ePub ISBN: 9781119972822
Mobi: 9781119972884

Set in 10/12 pt Times by Thomson Digital, Noida, India

Contents

The success of any application of NMR depends on the correct sample preparation, the appropriate use of parameters for data acquisition and processing; these are covered in Chapter 1. Once initial data has been collected to assess if a protein system is suitable for NMR studies, the next step will depend upon whether the objective is a determination of the three-dimensional structure of a protein or its complex, or a more limited specific objective such as screening for ligand binding or the determination of pK_a values. For all but the smallest proteins, isotope-labelling will be required (Chapter 2), and to go beyond purely qualitative experiments resonance assignments (Chapter 3) will be essential. Structure determination involves the acquisition and treatment of structural restraints (Chapter 4) and the use of these to obtain structural ensembles (Chapter 5). Chapter 6 describes the additional information on protein structure or complex formation which can be obtained when the protein contains a paramagnetic species – either naturally occurring or introduced specifically for the purpose. Chapter 7 describes different approaches to the study of the binding of small molecules, ranging from screening to full structure determination of the complex; this requires an understanding of the theoretical and practical aspects of the effects of chemical exchange in NMR, which is also important in many other areas of biological NMR. Chapter 8 provides a comprehensive description of the use of NMR to study macromolecular complexes; this is a challenging area and the chapter outlines the problems and approaches which can be taken to overcome these challenges. Chapter 9 focuses on the structural studies of intrinsically disordered proteins. The widespread existence and significance of these proteins are becoming increasingly recognised and NMR is currently the best method to provide detailed information on their conformational distributions.

NMR is uniquely suited for the characterisation of biomolecular dynamics. Since so many nuclei can be detected simultaneously, NMR can provide a comprehensive description of the internal motions and conformational fluctuations at atomic resolution, and NMR methods have been developed to quantify motions that occur at a wide range of timescales, from picoseconds to days and months. At the same time, consideration of dynamics and the averaging processes to which they lead is an essential part of the use of NMR to obtain structural information. As a result, several chapters in this book deal with methods for obtaining dynamic information from NMR. For additional information the reader is also directed to the following reviews [4–7].

Over the last few years there have been significant developments in the application of solid-state NMR techniques as a tool for determining the high-resolution structures of proteins, ranging from microcrystalline soluble proteins to protein fibrils and membrane proteins. It is now possible to assign the spectra of proteins larger than 100 amino acids using ^{13}C,^{15}N –labelling [8]. However, as yet this remains an area for the expert and it is not covered in detail in this book (although some of the methods for isotope-labelling described in Chapter 2 will also be relevant to solid-state studies). For useful reviews, the reader is directed to [9,10].

Note Added in Proof

Several valuable relevant reviews have appeared while this book was in production. In particular, two useful qualitative introductions to biomacromolecular NMR for the newcomer to the field would serve as valuable initial reading [11,12]. Clore [13] has reviewed the use of relaxation methods (see Chapters 6 and 7) to observe species with low population, Wishart [14] has reviewed the use of chemical shifts in structure determination (see Chapters 4 and 5), and Dominguez *et al.* [15] have reviewed the use of NMR in the study of protein-RNA complexes (see Chapter 8).

References

1. Cavanagh, J., Fairbrother, W.J., Palmer, A.G. III *et al.* (2007) *Protein NMR Spectroscopy: Principles and Practice*, 2nd edn, Academic Press, San Diego.
2. Keeler, J. (2005) *Understanding NMR Spectroscopy*, John Wiley & Sons, Ltd, Chichester.
3. Levitt, M.H. (2001) *Spin Dynamics: Basis of Nuclear Magnetic Resonance*, John Wiley & Sons, Ltd, Chichester.
4. Mittermaier, A.K. and Kay, L.E. (2009) Observing biological dynamics at atomic resolution using NMR. *TIBS*, **34**, 601–611.
5. Baldwin, A.J. and Kay, L.E. (2009) NMR spectroscopy brings invisible protein states into focus. *Nature Chem. Biol.*, **5**, 808–814.
6. Jarymowycz, V.A. and Stone, M.J. (2006) Fast time scale dynamics of protein backbones: NMR relaxation methods, applications, and functional consequences. *Chem. Revs.*, **106**, 1624–1671.
7. Igumenova, T.I., Frederick, K.K. and Wand, A.J. (2006) Characterization of the fast dynamics of protein amino acid side chains using NMR relaxation in solution. *Chem. Revs.*, **106**, 1672–1699.
8. Schuetz, A. *et al.* (2010) Protocols for the sequential solid-state NMR spectroscopic assignment of a uniformly labeled 25kDa protein: HET-s(1-227). *ChemBioChem.*, **11**, 1543–1551.
9. Renault, M., Cukkemane, A. and Baldus, M. (2010) Solid-state NMR spectroscopy on complex biomolecules. *Angew. Chem. Int. Ed.*, **49**, 8346–8357.
10. McDermott, A. (2009) Structure and dynamics of membrane proteins by magic angle spinning solid-state NMR. *Ann. Rev. Biophys.*, **38**, 385–403.
11. Kwan, A.H., Mobli, Gooley, P.R. *et al.* (2011) Macromolecular NMR spectroscopy for the non-spectroscopist. *FEBS Journal*, **278**, 687–703.
12. Bieri, M., Kwan, A.H., Mobli, M. *et al.* (2011) Macromolecular NMR spectroscopy for the non-spectroscopist: beyond macromolecular solution structure determination. *FEBS Journal*, **278**, 704–715.
13. Clore, G.M. (2011) Exploring sparsely populated states of macromolecules by diamagnetic and paramagnetic NMR relaxation. *Protein Sci.*, **20**, 229–246.
14. Wishart, D.S. (2011) Interpreting protein chemical shift data. *Prog. Nucl. Magn. Reson. Spectrosc.*, **58**, 1–61.
15. Dominguez, C., Schubert, M., Duss, O. *et al.* (2011) Structure determination and dynamics of protein–RNA complexes by NMR spectroscopy. *Prog. Nucl. Magn. Reson. Spectrosc.*, **58**, 62–87.

1

Sample Preparation, Data Collection and Processing

Frederick W. Muskett

Purgamentum init, exit purgamentum

1.1 Introduction

The power of NMR spectroscopy for the analysis of biological macromolecules is undisputed. During the last two decades, the development of spectrometers and the experiments they perform, software, and the molecular biological techniques for the expression and purification of proteins have progressed at a formidable rate. Enrichment of molecules in the three major isotopes used in NMR (^{15}N, ^{13}C and ^{2}H) is now commonplace and the cost is no longer prohibitive. The software used to analyse the plethora of data we can generate makes spectral assignment and the extraction of data straightforward for all but the most challenging systems. With all these developments it is easy to forget some of the more fundamental requirements for obtaining good quality NMR data, namely a good sample and a well-set-up NMR experiment.

1.2 Sample Preparation

The first, and possibly one of the most important, steps before embarking on an NMR-based project is the preparation of the sample. Spending some time optimising sample

Protein NMR Spectroscopy: Practical Techniques and Applications, First Edition.
Edited by Lu-Yun Lian and Gordon Roberts.

conditions for concentration, ionic strength, pH and temperature before collecting large amounts of data will pay dividends, particularly if the sample is difficult and expensive to produce. Ideally, the optimised sample will not only give the best possible NMR data, but will also have long-term stability as, assuming the project requires backbone and side-chain assignments, the total acquisition time required can be in the order of several weeks.

The following sections will outline the general requirements of a biological sample that is to be used to record NMR data. The assumption has been made that full resonance assignment is required; however, these guidelines could, and probably should, be applied to all samples regardless of the intention of the experiments.

1.2.1 Initial Considerations

This optimisation can be performed in the NMR spectrometer but much can be done using other biophysical techniques such as circular dichroism or fluorescence spectroscopy. These methods require much lower concentrations and do not require isotopic enrichment. The effects of buffer composition on secondary structure content and the melting temperature of the sample give a useful starting point. Once the initial conditions have been determined, final optimisation in the spectrometer can begin. If the sample is a protein, although much can be learned from a simple one-dimensional proton experiment, by far the most useful experiment is the ^{15}N-edited HSQC. This type of experiment removes a great deal of resonance overlap, allowing the user to see in much more detail the effects of varying pH, ionic strength and temperature.

NMR has intrinsically poor sensitivity and, as a result, the concentration of the sample needs to be in the millimolar range. For a conventional room temperature probe ideally the sample concentration needs to be $\geq$1 mM but can be as low as 0.5 mM. With the development of cryogenically cooled probes this concentration can be reduced to $\geq$0.2 mM and given the right sample can be as low as 0.05 mM (depending on the experiments performed). Whilst the sample can be exchanged into the buffer intended for NMR experiments in the last step of purification (usually gel filtration chromatography) it is rarely at the concentration required. The two main methods used to increase concentration are lyophilisation with subsequent re-suspension in a lower volume and ultra-filtration, or a combination of the two. Unfortunately, whether a sample will survive either method cannot be predicted; in the end one must simply try and see what happens. However, lyophilisation is generally considered the more dangerous of the two. The number and type of disposable ultra-filtration devices on the market is large, each with their own characteristics regarding compatibility with a particular sample and the effective volumes with which they can be used; again, try and see what happens. As a final step, either passing the sample though a 0.2 μm filter or centrifuging in a benchtop micro-centrifuge, to remove any insoluble material or dust, will greatly help sample homogeneity.

With modern solvent suppression techniques that can effectively eliminate the ~110 M protons from the water signal, dissolving the sample in ^2H_2O would, at first, no longer seem to be required. However, such samples are still important in recording the experiments designed to allow assignment of protein side-chain resonances and for ^{13}C-edited nuclear Overhauser effect (NOE) experiments. Even the most efficient solvent suppression techniques still leave residual solvent signal and at the same time suppress or distort the signals of interest in that area of the spectrum. In addition, the use of ^2H_2O allows one